ESPIONAGE BLACK BOOK SIX

In this series of technical monographs:

Espionage Black Book One: Intelligence Databases Explained

Espionage Black Book Two: Codes and Ciphers Explained

Espionage Black Book Three: Surveillance Explained

Espionage Black Book Four: Open-Source Intelligence Explained

Espionage Black Book Five: Basic Intelligence Explained

ESPIONAGE BLACK BOOK SIX:

Spy Tradecraft for Citizens Explained

Dennis B. Desmond
Troy Whitford
Henry W. Prunckun

Bibliologica Press

Disclaimer

The information provided in this book is for educational purposes only. Neither the publisher nor the authors are rendering advice to the reader. The ideas, suggestions, and procedures provided in this book are based on academic study and are not intended as a substitute for seeking professional guidance or legal advice. This book is distributed on the understanding that neither the publisher nor the authors or copyright owner shall be held responsible for any loss or damage or held civilly or criminal liable for any alleged act by the reader arising from any suggestion or information contained in this book.

Espionage Black Book Six:
Spy Tradecraft for Citizens Explained

by Dennis B. Desmond, Troy Whitford, and Henry W. Prunckun

Copyright © 2022 by Henry W. Prunckun

ISBN 978-0-6456209-0-0

A catalogue record for this book is available from the National Library of Australia

For information on all Bibliologica Press's publications, visit our Web site at bibliologica.com

Bibliologica Press
P.O. Box 656
Unley, South Australia, 5061
Australia

CONTENTS

— CHAPTER ONE —

WHAT IS TRADECRAFT?

W hen we think of *tradecraft*, we are likely to imagine some television or cinema film about daring-doo adventures by legendary fictional characters— James Bond, Jason Bourne, Paul Deveraux, and Jack Ryan. Take, for instance, the opening scene of *Thunderball*.[1] It shows Bond secretly observing the funeral of enemy agent Colonel Bouvar. There is only one mourner present—a respectably dressed blonde-haired woman in a black dress and veil.

So, why is Bond interested in her? Why does he wait for her to return to her opulent manor? Viewers then watch Bond confront the woman, whom we presume to be the widow, by punching her!

But *she* is *he*, and he is the dead Colonel in disguise.

How did Bond know this? *Spy tradecraft*, of course!

The entertainment industry has conditioned us to set aside disbelief as we are led to think that spies can do impossible things. We have been presented with hundreds of movies and countless novels about how espionage operatives use tradecraft to accomplish extraordinary feats. Although based on fact, the stories are pure fabrication. The standard disclaimer in novels states: "This is a work of fiction. The names, characters,

---

1. Based on Ian Fleming's book, *Thunderball* (London: Jonathan Cape, 1961).

businesses, events, and incidents are the products of the author's imagination. Any resemblance to actual persons, living or dead, or actual events is purely coincidental."

Returning to *Thunderball*, Bond's uncanny deductive achievement (or was it clairvoyance[2]) was only capped by his escape from the Colonel's estate—via a jetpack that rocketed him to the safety of his spycraft-infused Aston-Martin. It is easy to argue that these depictions have influenced cinema-going audiences that intelligence agencies can see all and do all, in total secrecy, using spy tradecraft.

So, what is tradecraft? It is the art and science of spying. The word *craft* refers to making things, but in intelligence work, craft refers to the tactics, techniques, and procedures (TTPs)[3] used by spies to gather information. In this sense, spies use methods that have proven successful over centuries to conceal who they are and their collection activities. In the fifth century BCE, the Chinese scholar Sun Tzu posited that "foreknowledge enables the wise sovereign and the good general to strike and conquer, and achieve things beyond the reach of

---

2. On the topic of "fantastic," as a general observation about the spy genre is that writers seem to think that it is necessary to write into the plot *more*—more of everything—guns, fights, chases, explosions, and bodies. Writers seem to think that these excesses make for better reading, but what it does is shift their storytelling from fiction into the realm of fantasy.

3. According to the *Department of Defense Dictionary of Military and Associated Terms* (Washington, DC: DoD, 2016), *tactics* refers to the "employment and ordered arrangement of forces in relation to each other." *Techniques* means "non-prescriptive ways or methods used to perform missions, functions, or tasks." And *procedures* refer to "standard, detailed steps that prescribe how to perform specific tasks." pp. 235, 239, and 190, respectively.

ordinary men."[4]   We now refer to foreknowledge as *intelligence*.

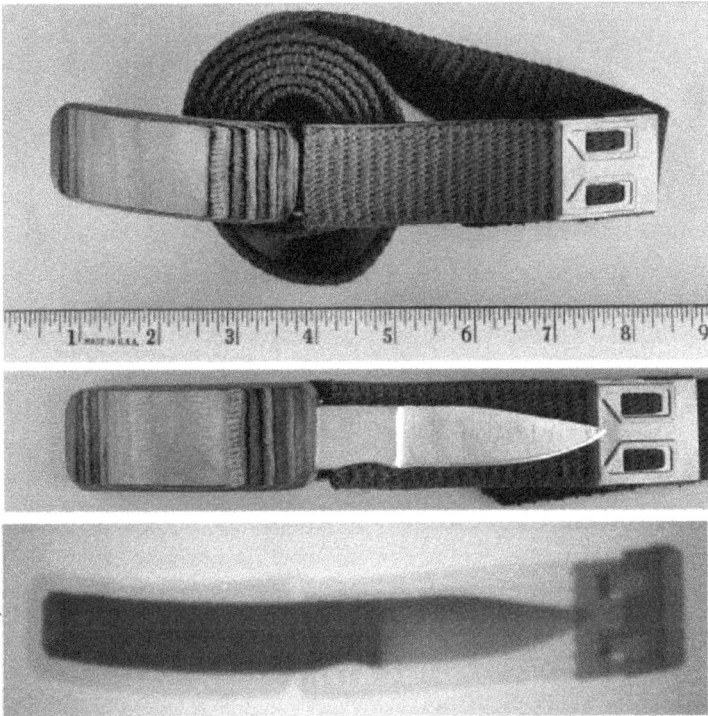

Figure 1—This knife concealed in a belt is an example of a specially designed device. Courtesy of the FBI.

Spies use devices to assist them in their undercover tasks, termed *spycraft*.[5] These devices are also considered to be part of the craft. Such equipment can be everyday items (improvised) or specially designed apparatuses (planned).

---

4. Sun Tzu, translated by Lionel Giles (1910), *The Art of War* (Mumbai: Wilco Publishing House, 2009), p. 110.

5. Robert Wallace and H. Keith Melton, with Henry Robert Schlesinger, *Spycraft: The Secret History of the CIA's Spytechs from Communism to Al-Qaeda* (New York: Dutton, 2008).

Tradecraft also refers to the techniques, tactics, and procedures used by counterspies—those tasked with catching spies. This category of spying is termed *counterintelligence*. Counterintelligence officers use additional techniques, tactics, and procedures to identify people trying to hide their identity and their involvement in spying.

Tradecraft comprises a suite of methods that form the bedrock for all intelligence operations. This is because, at its simplest, intelligence is research, but some aspects are secret.[6] This could be the fact that research is being conducted, what information is being collected, how the information is being collected and by whom, how it is collated with other data and analyzed, and the level of certainty attributed to the research findings.

Therefore, tradecraft, supported by spycraft, helps ensure that those aspects of an intelligence operation that need to be protected will be and they will remain secret long after the operation is finished.

It follows, then, that a citizen who needs to do something to keep themselves safe can adopt these methods. These TTPs can also be used to protect other people and the things they hold dear. In the next chapter, we will underscore this point where we discuss why a citizen could be interested in using spy tradecraft.

Our discussion of tradecraft might lead some readers to think that the authors have divulged state secrets, endangering national security by revealing espionage techniques. We have not. The information we present is in the public domain *if* you know where to look for it *and*

---

6. Hank Prunckun, *Methods of Inquiry for Intelligence Analysis, Third Edition* (Lanham, MD: Rowman & Littlefield, 2019), p. 4.

can distinguish between fact and fiction, or in some cases, fantasy.

Nevertheless, we acknowledge that there are some operational aspects of tradecraft that we have kept to ourselves because we have sworn under various official government secrecy laws to protect this information, as well as, in some cases, given our promise to keep confidential private client contracts. We have practiced secrecy in operational careers because it is the foundation of all intelligence work.[7]

What we have done is to take a sensible approach to provide citizens with an understanding of the theory and practice of tradecraft. We have applied our collective operational experience in addition to our academic training to separate truth from fabrication and to distil the voluminous amount of information on spies and spying to provide you, the reader, with an explanation of spy tradecraft.

Whether you are travelling through or living in troubled lands, you are the potential target of criminals, sociopaths, or just what to live a quiet life; this book will help guide you. Is this book a "complete book of..." or a "comprehensive manual of..."? No, it is not. As we said before, it provides an *explanation* of spy tradecraft for citizens.[8]

What are the tactics, techniques, and procedures that a citizen needs to consider? We have extracted the

---

7. Hank Prunckun, *Counterintelligence Theory and Practice, Second Edition* (Lanham, MD: Rowman & Littlefield, 2019), p. 37.

8. And the information provided in this book is for educational purposes only. See the "Disclaimer" on page iv.

essentials and developed a theory that comprises four premises:[9]

1. Secrecy in planning;
2. Being able to move surreptitiously;
3. Anticipating hostile intentions to be able to frustrate these schemes; and
4. Accomplishing your plans by the "simplest and least intrusive means possible."[10]

The chapters of this book follow the four premises of theory because, without an understanding of tradecraft, everything a citizen does could potentially expose them to those who have ill intent. Now, let's turn our attention to why you are interested in intelligence tradecraft.

---

9. The first three premises of our theory are grounded in the tactical thinking of Sun Tzu, *The Art of War* (2009), pp. 27–28.

10. This fourth premise was adapted from Peter Earnest, in Jack Barth, *International Spy Museum Handbook of Practical Spying* (Washington, DC: National Geographic, 2004), p. 13.

— CHAPTER TWO —

WHY USE TRADECRAFT?

History shows that the Internet inventors envisioned a global communications platform with unlimited information to provide a digital ecosystem of equality and egalitarianism.[11] Yet, it was not too long before the Internet, and the associated World Wide Web,[12] became a platform for criminal activity, a channel for misinformation and hate speech, and a surveillance platform used by autocratic governments and fraudsters.[13]

The Internet also enabled the rise of the so-called *surveillance economy*, creating a nexus between the digital and physical world, which allowed information collectors to collect large amounts of personal, sensitive, and behavioral data. Many have turned to tactics, techniques, and procedures (tradecraft) and used specialized software applications and devices (spycraft) to circumvent on-line and physical collections.[14]

---

11. Janet Abbate, *Inventing the Internet* (Cambridge, MA: MIT Press, 2000).

12. The World Wide Web, or the Web for short, are the pages that are delivered to your browser when you are online. In contrast, the Internet comprises the physical network of routers, switches, modems, servers, and computers where these Web pages reside and emails and files traverse.

13. Jonathan Clough, *Principles of Cybercrime* (Cambridge: Cambridge University Press, 2015).

14. H. Keith Melton; Henry Robert Schlesinger; and Robert Wallace, *Spycraft* (New York: Random House, 2011).

The rise of populist movements and the growth of extremism, global economic instability resulting from COVID-19, and political upheavals in places like Hong Kong[15] have resulted in laws restricting individual freedoms and increasing surveillance by governments and private entities.

These legal changes have affected the rights of individuals driven to seek methods to protect their privacy and prevent the exploitation of their identity data.[16] A natural alternative for many has been to adopt the tradecraft techniques and spycraft devices used by those who work undercover to engage in lawful activities without fear of persecution.

The term *persecution* is not used lightly; it is a fact of life in some countries, like Iran. In September 2022, the Iranian authorities arrested Mahsa Amini, who failed to comply with a headscarf law. She died in police custody, and her fellow citizens protested her death. As the Iranian authorities tried to stifle these public displays, the U.S. Treasury Department lifted some previous restrictions that were placed on Iranian Internet usage. This was done to help citizens communicate with the world about Iran's violent crackdown on peaceful protests alleging police brutality.[17]

---

15. Au Loong Yu, *Hong Kong in Revolt: The Protest Movement and the Future of China* (London: Pluto Press, 2020).

16. Dennis B. Desmond, "After Roe v Wade, Here's How Women Could Adopt 'Spycraft' to Avoid Tracking and Prosecution," in *The Conversation*, June 30, 2022. Available at: https://theconversation.com/after-roe-v-wade-heres-how-women-could-adopt-spycraft-to-avoid-tracking-and-prosecution-186046. Accessed: June 30, 2022.

17. Office of Foreign Assets Control, *Iranian Transactions and Sanctions Regulations, 31 CFR part 560, General License D-2,*

There may be other reasons to employ tradecraft and spycraft: concern for your sensitive data so that you can increase or maintain your privacy or pursue personal and lifestyle choices. These reasons may seem trite but consider the oppressive political atmosphere in Vladimir Putin's Russia following that country's 2022 illegal war against Ukraine.

While it is true that extremists and criminals may employ these same tactics, techniques, and procedures (TTPs), it is essential to afford the same level of protection for potential victims of stalking, religious persecution, domestic abuse, exploitation, or any number of other oppressive acts.[18]

At its heart, tradecraft is about *identity management.* While governments can spend millions on training their intelligence and law enforcement officers in techniques to protect their identities, citizens are limited by not having enough knowledge or the funds to do what is needed to defend themselves. However, you may use some tactics, techniques, and procedures to protect your personal information from being collected and exploited by those who would harm you.

Applying TTPs can assist with information control and security while reducing public exposure. Viewed another way, spy tradecraft is about managing routine behaviours

---

General License with Respect to Certain Services, Software, and Hardware Incident to Communications (Washington, DC: U.S. Department of the Treasury, 2022).

18. Dennis B. Desmond; P. Salmon; and D. Lacey, "Functional Systems within Cryptolaundering Processes: A Work Domain Analysis Model of Cryptolaundering Activities," in *Journal of Cyber Policy*, 2021, Volume 6, Number 2, pp. 155–176.

and activities, increasing your situational awareness,[19] and applying specific techniques to control the release, exposure and sharing of your identity data. Spycraft is about using technologies to secure your communications, conceal your activities, and make discovery and detection more difficult by outside observers. Physically, tradecraft allows citizens to engage in activities, often in plain sight.

> While serving as a US Army and FBI special agent, a human intelligence support officer, and a senior counterintelligence officer for DIA, I have had to practice and recognize the use of tradecraft. Intelligence and law enforcement officers employ tradecraft to protect assets, support intelligence collection activities, and protect the lives of clandestine collectors and special operators. Tradecraft has evolved from purely physical activity to a hybrid combination of physical and digital activities to ensure the successful execution of law enforcement and intelligence activities.
>
> *Dennis B. Desmond, 2022*

Tradecraft also extends to the on-line environment. On-line tradecraft achieves the same risk mitigation by using time-tested TTPs—password management, using virtual private network applications or using "The Onion Router,"—in short, known as *Tor*[20]—properly configuring

---

19. Albert A. Nofi, *Defining and Measuring Shared Situational Awareness* (Alexandria, VA: Center for Naval Analyses, 2000).

20. The Onion Router is a secure Web browser that enables anonymous communication. Tor is constructed to allow citizens to conceal their log-on location and makes it more difficult to trace a citizen's Internet activity. In brief, Tor protects citizens' privacy, thus providing them with the freedom to communicate confidentially. Nitul Dutta; Nilesh Jadav; Sudeep Tanwar; Hiren Kumar Deva Sarma; and Emil Pricop, "Tor—The Onion Router," in

your browser's privacy settings, and using applications to delete Internet search history and remove tracking cookies and web bugs.

Even the best surveillance operatives can avoid being detected. No matter how good a surveillance operative is, a well-trained and practiced spy can successfully communicate with an agent-in-place or a "handler" to pass information. And so can citizens.

Dennis B. Desmond, 2022

National security, military, law enforcement, and private intelligence agencies, as well as private investigators and open-source researchers, conduct *identity operations*. These operations can either expose or obscure identity data by using offensive or defensive TTPs, respectively. That is, offensive "ops" expose information that is being kept confidential, and defensive ops try to protect data.

By way of example, national security and law enforcement intelligence agencies use offensive ops to target terrorists, organized criminals, and foreign spies. Offensive tactics, techniques, and procedures include identity intelligence, identity resolution, identity exploitation, de-anonymization and re-identification, forensics (medical and device) and sensitive site exploitation.

In contrast, defensive identity TTPs are designed to conceal or obscure (by, say, inserting or removing) identity data into Web pages or on-line databases to prevent its collection or lead surveillants off-track. Defensive identity operations can be scaled up to include

---

Cyber Security: Issues and Current Trends. Studies in Computational Intelligence, (Singapore: Springer, 2002), Volume 995.

sophisticated countermeasures that intelligence agencies term *identity and persona management*. This may involve removing or suppressing a person's identity (think witness protection), counter-forensics, and counter-surveillance and counter-collection activities.

Every interaction you have with another entity, such as a person, a company, an electronic device, a networked system, or a software application, can be considered an *identity transaction*.[21] Each transaction generates information about who you are from your interaction with the entity. These data can be collated, analyzed, and then sold.[22] While commercial entities may use these data to verify who you are (as in the case of fraud prevention), to conduct creditworthiness checks or marketing, it provides *basic intelligence*[23] to national security and law enforcement intelligence agencies.

---

21. U.S. Department of Defense, *Joint Doctrine Note 2-16, Identity Activities* (Washington, DC: U.S. Department of Defense, 2016), pp. I-11–I-14.

22. As an example, take the case of Amazon's bid to merge with the Roomba vacuum maker iRobot. Privacy critics warned that another vector by which companies could gather personal data, in this case, floor plans of your homes. *ABC News*, "Amazon's Roomba Robot Vacuum Merger Bid Sparkes Privacy Fears." Available at: https://www.abc.net.au/news/2022-08-28/amazon-rumba-bid-sparks-privacy-fears/101374130 Accessed: September 16, 2022.

23. "Basic intelligence" is "…factual information which results from the collection of encyclopedic information of more or less permanent or static nature and general interest which, as a result of evaluation and interpretation, is determined to be the best available. Chief of Naval Education and Training, prepared by Terry L. Schroeder, *Intelligence Specialist 3 & 2, Volume 1* (Washington, DC: U.S. Government Printing Office, 1983), p. AIII-15.

For example, when a customer purchases goods or services using a credit card or ATM card, the retailer will attach what is called a *persistent identifier*, or PID for short.[24] These PIDs as they are referred to are like the PIDs libraries and archives use to identify books and documents.[25] However, in the case of consumer transaction PIDs, they are used to perform *identity resolution*.

Each identity transaction is attached to a PID allocated to you and is enhanced when you present your loyalty card or your e-mail address. Suppose you also provide your telephone number, address, or zip/postcode. In that case, the retailer can use it to gauge your solvency, entertainment preferences, leisure activities, on-line and physical shopping behaviors, food, music and reading preferences, and more.[26]

How can they do this? By cross-checking your PID data against public records such as civil court judgments, telephone listings, residential tenancies records, land sale records, consumer credit bureau databases, and data purchased from on-line data collectors—large Web-based retailers, for example.[27] Information you post to social

24. John A. Kunze, *Towards Electronic Persistence Using ARK Identifiers* (Oakland, CA: California Digital Library, 2003), Section 3.

25. See, Open Researcher and Contributor ID (ORCID) found at: https://orcid.org/. Accessed September 19, 2022.

26. Jeff Jonas, "Identity Resolution: 23 Years of Practical Experience and Observations at Scale," in *Special Interest Group on Management of Data, Association of Computing Machinery, International Conference on Management of Data*, June 27, 2006.

27. Susan Landau, "FTC [U.S. Federal Trade Commission] Lawsuit Spotlights a Major Privacy Risk: From Call Records to Sensors, Your Phone Reveals More About You Than You Think,"

media sites such as *Facebook*, *Instagram*, and *Twitter* are understood to be aggregated, analyzed, and also attached to your PID.

Some businesses specialize in collecting images people post and attaching metadata to them, again, to identify the individuals. This is a legitimate business endeavor that allows the images to be used by law enforcement to identify victims and suspects or by retailers to "know their customers" and immigration and customs officials. Nevertheless, these same data can also be used for nefarious purposes by criminal elements to, say, coerce you to pay them for their removal from public view.

Don't forget that these images allow miscreants to stalk you. Criminals leverage social media data to manipulate people into providing access to their financial accounts or use this information to represent themselves as you to commit fraud (i.e., identity theft).[28]

A less apparent means of data collection occurs through your Internet-connected devices, your communications with friends and family, and your interactions with Web retailers. Information about your physical location, habits, activities and even the devices and applications you use are collected, aggregated, and attached to your PID. Cookies deposited by websites to your hard drive allow businesses to record your preferences and verify whether you are a repeat visitor.

---

in *The Conversation*, published on-line August 22, 2022. Available at: https://theconversation.com/ftc-lawsuit-spotlights-a-major-privacy-risk-from-call-records-to-sensors-your-phone-reveals-more-about-you-than-you-think-189618. Accessed: September 17, 2022.

28. Jonathan Clough, *Principles of Cybercrime*, 2015.

Used benevolently, your data can enhance your on-line experience and provide you with useful help and guidance. Moreover, used lawfully, this information helps law enforcement and intelligence agencies to identify criminal activities. Unfortunately, your digital data can also be used to defraud you or victimize you because of your political, social, or religious beliefs or affiliations. This type of data misuse often happens through *doxxing*.

Doxxing (also spelled *doxing*) is collecting public and private information about an individual and publishing that information on-line to embarrass, harass, punish, or intimidate a person.[29] Information sought after includes, but is not limited to:

- Bank account details;
- Civil court judgments;
- Confidential photos;
- Credit card details;
- Criminal records;
- Distressing personal events;
- Employment details;
- Private e-mails;
- Residential address;
- Social security number; and
- Telephone number.

---

29. Although the practice of revealing a person's details without permission predates the World Wide Web, the term doxxing was coined in the 1990s when disputes between rival hackers would result in one of the parties "dropping docs" on the other. "Docs," short for documents, evolved to become "dox," and from there, it was a short step to make the word a verb, but without the preface "drop."

Publicly available information (PAI) can be used by stalkers, criminals, and politically or religiously motivated extremists to target individuals by publishing the results on-line.[30] Take, for example, the infamous hacktivist group *Anonymous*,[31] which, in 2011, doxxed hundreds of confidential U.S. law enforcement personnel details in retaliation for what it claimed were the arrests of fellow hackers. These data included "...private e-mails, passwords, training files, data from informants, Social Security Numbers, and stolen credit card information from an on-line sheriff's store."[32]

Doxxers typically access this information through social media sites, identity retailers and brokers (i.e., so-called "people finders"), or using pretext[33] tactics. Doxxers may encourage other on-line actors to use the data to target and victimize their "marks." The outcome can be devastating when doxxers mistake their marks for

---

30. Jeffrey Pittman, "Privacy in the Age of Doxxing," in *Southern Journal of Business and Ethics*, 2018, Volume 10, pp. 53–58.

31. Gabriella Coleman, *Hacker, Hoaxer, Whistleblower, Spy: The Many Faces of Anonymous* (New York: Verso Books, 2014).

32. Elinor Mills, "AntiSec Hackers Post Stolen Police Data as Revenge for Arrests," in *CNET News*. Available at: https://www.cnet.com/news/privacy/antisec-hackers-post-stolen-police-data-as-revenge-for-arrests/ Accessed, August 29, 2022.

33. "Pretext should not be confused with the term *social engineering*, which has gained popularity in recent years. Social engineering is a slang term that commonly refers to the individual act of manipulation (usually for fraudulent purposes) to gain unauthorized access to IT systems. This is vastly different from its true meaning, which is large scale societal planning. So, the use of the term social engineering in the context of accessing information surreptitiously is incorrect." Hank Prunckun, *Scientific Methods of Inquiry for Intelligence Analysis, Second Edition* (Lanham, MD: Rowman & Littlefield, 2015), p. 153.

another person, especially when falsely accused of a crime.[34] While celebrities and politicians are often the marks of doxxing, doctors, law enforcement officers, military family members, and political activists are also at risk.

"The art of using pretexts is a science and should be approached as one."

Greg Hauser, *The Pretext Manual* (Austin, TX: Thomas Investigative Publications, 1994), p. 5.

One approach to defending your identity is the use of *segmented identities*; you can do this by creating several on-line personas. The identity segmentation approach means each persona has an e-mail account, social media profile, and on-line shopping accounts.

Persona profiles include an alias name, profile image, and biography with an attached e-mail account such as *Proton Mail* or using the *Virtru* extension for *Gmail*. Each persona may have its social media profile with its community of interest, segregating your persona's interests and connections to avoid aggregation—the aggregation of information helps builds what could be considered a "dossier" on you. Perhaps the most challenging part of this process is learning that it's okay not to tell the truth to protect your privacy and obscure your identity.[35]

---

34. Kyle Quinn was wrongly identified as a participant in an extremist demonstration in Charlottesville, N.C. and doxxed, resulting in harassment and threats. Jeffrey Pittman, "Privacy in the Age of Doxxing," 2018.

35. It is NOT okay to provide false information to a law enforcement, regulatory, or compliance officer, a judicial officer

17

> There are companies that profit from the collection and sale of your identity data. Check the company's website options for removal and suppression of your identity-related information that they are selling.
>
> *Dennis B. Desmond, 2022*

Identity segmentation also includes setting-up browser profiles for each persona. It is critical to ensure your answers to security questions, and biographical data do not match your true personally identifiable information (PII).

Segmentation will also require you to employ spy tradecraft skills, such as using only stored value cards (SVCs) instead of using true-name debit and credit cards on-line to purchase and ship goods to a post office box for security. Segmenting your identities into personas is best supported by using multiple cell phones with pay-as-you-go SIM cards, all paid for with cash cards and only turned on when necessary.

To communicate securely with contacts, we suggest you use encrypted applications to protect your files, folders, and storage drives. Ensure that websites you visit have encryption enabled. What does this mean? You may have noticed that some URLs start with https:// rather than

---

or in any circumstance where you are being asked, by someone with legal authority, to be honest. It is also NOT okay to defame someone. Defamation is where a person says (which is known as *libel*) or writes something that harms the reputation of a third person. As a rule-of-thumb, you can provide invented details about yourself as long as it is not for fraudulent purposes. For instances, writers often write under a pseudonym; actors perform using "stage names"; and, we all have told "white lies" to avoid hurting people's feelings, e.g., "Your carrot cake is so moist, Aunt Penelope."

with http://. The additional "s" in the URL means that your website connection is secure using encryption.

Figure 2—Segmenting your identities can be supported by several cell phones with pay-as-you-go SIM cards.

In practice, any information you send to the website is safe from *interception*.[36] It is not safe from data retention laws—see the discussion in the section "Premise Three," in Chapter Six.

In some cases, where you wish to communicate securely, you will find that applications such as *Telegram*, *Signal*, *WhatsApp*, or *Facebook Messenger*—all with encryption enabled—are helpful. Applying the segmented identity approach will provide identity protection against identity theft and fraud and delay or prevent identity resolution efforts by extremists and criminals.

---

36. The "s" designation refers to *Secure Sockets Layer*, or SSL for short. This is done through an SSL *certificate*. An SSL certificate is a small data file that initiates a cryptographical link between the Web server and your computing device. Once the link is encrypted, it ensures that the information exchanged between your device and the Web server are private.

— CHAPTER THREE —

PREMISE ONE—SECRECY IN PLANNING

C ommunication is an essential part of everyday life. Communication is the act of sending a message to a receiver who can understand it.[37] Citizens use various methods to do this, from face-to-face discussions to electronic devices that allow people world-wide to exchange information.

SECRET COMMUNICATION

Communication allows us to do simple things—like announcing the appointment of a new dentist in the district, to complicated tasks, such as instructing student chemists to conduct a laboratory experiment. We can think of many more examples, but what these acts of communication share is that the communication is open; they are not confidential, and they do not require the transmission of the message from sender to receiver to be secret.

In an era gone by, this might have been accomplished by two people discussing an issue. If neither person passed on the nature and content of this discussion to others, the information was, arguably, one hundred per cent secret.

---

37. Göran Sonesson, "Translation and Other Acts of Meaning: In Between Cognitive Semiotics and Semiotics of Culture," in *Cognitive Semiotics*, Volume 7, Number 2, 2014, pp. 249–280.

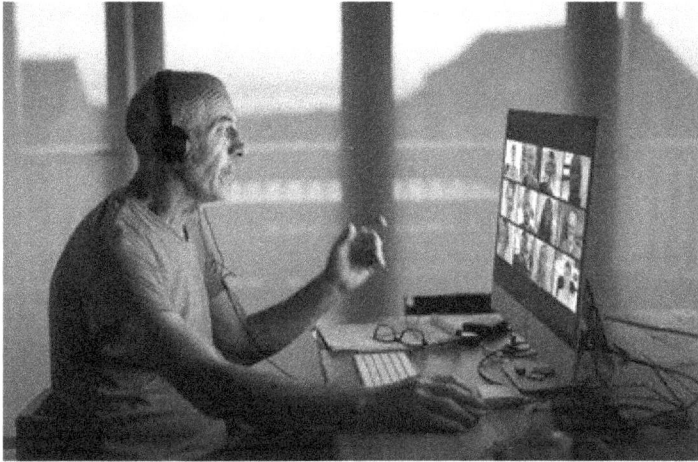

Figure 3—Worldwide audio-visual communication via a computer-based application.

Now enter modern society. Two people meeting to discuss a matter can be overheard.[38] Even if the conversation can't be picked up by someone close by, the fact that the two met can imply something is being planned. If we introduce an electronic device into the situation—say, a cell phone text or an e-mail—we will, of course, have eliminated the possibility of someone overhearing the conversation. Still, it does not mean that the communication is private.

Every method of communicating, despite its efficiency, has weaknesses that allow plans and schemes to be intercepted by parties not privy to them. It is a fact of life that people need to keep certain pieces of information

---

38. Scott R. French, edited by Margaret McFadden French, *The Big Brother Game* (Secaucus, NY: Lyle Stuart, 1975); Mick Tyner, *Surveillance Audio Amplifiers: The Cult of Super-Hearing* (n.p.: Trentland Press, 1990); and, Mick Tyner, *The Spook Book: A Strange and Dangerous Look at Forbidden Technology* (n.p.: Trentland Press, 1989).

private. We all have heard the words, "He doesn't talk about 'that'," with reference to, perhaps, some traumatic event in the person's life. Clearly, people do not want to communicate details of some personal issues with anyone.[39] To safeguard this information, they keep it to themselves. However, there are other pieces of information that people need to share but only with a select person or people, but no others.

Guarding one's communications comprises two factors—privacy and security. Privacy is the ability to keep information about oneself secluded, as well as the ability to communicate personal information as one chooses.[40]

Privacy implies that some aspect of the information they hold or what they are doing is sensitive. This is where the allied concept of security comes in. Security encompasses tactics, techniques, and procedures that help protect these sensation items of information.

Privacy and security are legal concepts that have been enshrined in constitutions, legal statutes, and case law of many liberal democracies. These laws are intended to steer those exercising executive powers away from unauthorized searches and seizures of personal information, commonly termed *invasion of privacy*.

The key to understanding this legal concept is the word *unauthorized*. Of course, governments can obtain warrants, and corporations and private individuals to apply for court orders (e.g., subpoenas) that compel

---

39. e.g., Political, and religious thoughts, health concerns, sexual orientation, past life events, or future aspirations.

40. It also relates to keeping oneself secluded if one chooses.

people to turn over specified items. In this way, the courts act as independent umpires of society's grant of privacy.

The issue comes when legislators pass laws that are immoral, unconstitutional,[41] or, as in some countries like North Korea, tyrannical. Even in the liberal West, there are examples. Take, for instance, the so-called "Jim Crow Laws"[42] that existed in the late-1800s and throughout the first half of the 1900s. These laws enforced racial segregation in the southern states of the United States. Just because a law requires someone to do something or prevents them from doing it does not mean that it should be obeyed. History is replete with other examples of such unjust laws.

It is about these types of unconscionable laws that we discuss secret planning and private communication. We are not writing for those engaged in, or are planning to commit, *mala in se* laws. These laws cover acts that society considers inherently wrong, such as theft, assault, rape, kidnapping, torture, murder, or a crime against humanity.

One may well ask: 'How can you advocate breaking some laws and obeying others?' The answer lies in the fact that there are two types of laws: just and unjust. I would be the first to advocate obeying just laws. One has not only a legal but a moral responsibility to obey just

---

41. See the writings of Thomas Aquinas on the topic of law, tyranny, and resistance. In short, he advocated that when those in legal authority exceed their mandate (e.g., pass repressive laws), citizens are free to disobey. Nico Patrick Swartz, "Thomas Aquinas: On Law, Tyranny and Resistance," in *Acta Theologica*, Volume 30, Number 1, (June 2010), pp. 145–157.

42. C. Vann Woodward, with an Afterword by William S. McFeely, *The Strange Career of Jim Crow* (New York: Oxford University Press, 2002). Originally published in 1955.

laws. Conversely, one has a moral responsibility to disobey unjust laws. I would agree with St. Augustine that an unjust law is no law at all.'[43]

In planning to perform an act, or to refuse to comply with some action that is required by an unconscionable law, several forms of communication could be used. The first that present themselves is speaking to another person face-to-face, talking to people via a telephone, sending a written message either on paper or through an electronic carrier (e.g., e-mail, text, or social media), or indirectly, say, through a third party. Recall communication also takes place with yourself, such as making reminder notes, stepwise instructions to follow, checklists, confirmation of appointment dates and times, and so on.

Although intelligence agencies do not reveal the types of safeguards they use to protect their communications when planning an operation, we know from declassified documents, manuals, and firsthand accounts by former operatives what these tactics, techniques and procedures involve. We also have a clear idea of the types of devices that are used to facilitate secret communications. However, we may not know what these devices look like or their exact technical specifications.

Take, as an example, mobile telephony—cell phones. We know the range of cell phones on the market and their specifications. But these are not the same phones that intelligence officers use in the field. Why? Because the

---

43. Martin Luther King, Jr., *Letter from a Birmingham Jail*, April 16, 1963.

information they are exchanging carries a higher *risk*. Risk is a factor of *consequence* and *likelihood*.[44]

The greater the consequence of the information becoming known to unauthorized parties, the higher the risk. But the consequence of something happening needs to be assessed in relation to the likelihood. Even if the likelihood is near zero ("rare"), the risk may still be "high." Table 1 shows the relationship between consequence and likelihood and the descriptors used to categorize risk.

$$risk = likelihood + consequence$$

Risk assessments are conducted in relation to a variety of situations; you may have been part of a committee at work to assess the risk of harm in the workplace. In our context, risk can be applied to plans becoming known to third parties before, during, and after the event.

Table 1—A typical risk assessment matrix.

	Consequences				
	1	2	3	4	5
Likelihood	Insignificant	Minor	Moderate	Major	Catastrophic
A Almost Certain	Moderate	High	Extreme	Extreme	Extreme
B Likely	Moderate	High	High	Extreme	Extreme
C Possible	Low	Moderate	High	Extreme	Extreme
D Unlikely	Low	Low	Moderate	High	Extreme
E Rare	Low	Low	Moderate	High	High

Depending on the activity, you need to decide which of these three stages there is an unacceptable risk so that you can "treat" the risk. If the risk is within your tolerance

---

44. The term *probability* is sometimes used instead of *likelihood*. Both terms are acceptable.

(i.e., if you can bear the consequences), you can accept the risk and proceed. If the risk is not tolerable, then you need to put in place things to mitigate the risk.

Citizens can implement simple but effective mitigation measures[45] for different modes of communication-based on the type of *threat* they face.[46]

We have listed below what we consider to be three generic sources of threat to a citizen who is trying to keep quiet about their plans. This list is intended to give you an appreciation of the hierarchy of threats you might face.

Steps taken to thwart intelligence collection at, for example, Level 2 would be sufficient to guard against any attempt by the lesser Level 3 threat, but not the reverse. This is important to remember. Citizens must determine where their threats lay before deciding on the range and depth of security measures they will require. Moreover, a citizen's threat level may change from time to time because of the dynamics of their operation. Therefore, a person's security needs will also be required to escalate or abate in response to these changing conditions.

Level 1 Threat—Surveillance by a foreign government's security intelligence agency or surveillance by one's own national law enforcement, regulatory, or

---

45. In intelligence parlance, mitigation measures are referred to as *countermeasures*.

46. "A *threat* is a person's resolve to inflict harm on another it is important to note that a threat cannot be posed by a force of nature or a natural event—these are *hazards*. Only people can pose a threat because they need *intent* and *capability*." Hank Prunckun, *Methods of Inquiry for Intelligence Analysis, Third Edition* (Lanham, MD: Rowman & Littlefield, 2019), p. 181.

compliance agency acting on their statutory powers or with a court-ordered warrant or subpoena.

Level 2 Threat—Surveillance by a state or local law enforcement or intelligence unit, an organized criminal group, a foreign or domestic business competitor employing a "spy-for-hire," a private detective acting on behalf of a party interested in your affairs, or other professional fact finders (e.g., investigative journalist).

Level 3 Threat—Non-professional surveillance, for example, by a jealous (ex-)spouse, stalker, nosey neighbor, envious co-worker, disgruntled employee, an untrustworthy business associate or competitor, or another interested individual or group acting on their own for ideology, profit, vexation, or revenge.

Let us look at a few examples of communication countermeasures (i.e., steps to mitigate surveillance at each threat level).

Communication is a part of living in the age. But living in this age comes with risks that private communications can become known to other parties. These communications are confidential, and they require the transmission of the message from sender to receiver to be secret to "Protect your home, your children, your assets, and your life."[47] We have examined a few ways this can happen in the electronic world, but in the next chapter, we will look at how confidentially can be preserved in the physical world.

---

47. J.J. Luna, *How to be Invisible, Third Edition* (New York: Thomas Dunne Books, 2012), p. dustjacket cover.

— CHAPTER FOUR —

PREMISE TWO—MOVING SURREPTITIOUSLY

M ovement around cities or regional areas is almost continuously monitored by static surveillance cameras. Used predominately to protect infrastructure and property, increasingly fixed surveillance cameras are also being used to monitor our daily movements, routines, and habits. Whether used under the pretense of developing "smart cities"[48] deterring crime, or as overt as the Chinese Communist Party's *social credit* system,[49] it is unlikely anyone can move throughout a town or city without appearing on at least one static camera.

COVER AND DISGUISE

In the current technological age, undetected movement is a challenge, but the answer may lay in pre-technological spy tradecraft such as *cover* and *disguise*. Adopting these techniques to avoid detection is probably as old as the spying profession. Still, it is likely to provide the best

---

48. Trevor Braun; Benjamin C.M. Fung; Farkhund Iqbal; and Babar Shah, "Security and Privacy Challenges in Smart Cities," in *Sustainable Cities and Society*, 2018, Volume 39, pp. 499–507.

49. Qiang, Xiao, "The Road to Digital Unfreedom: President Xi's Surveillance State," in *Journal of Democracy*, 2019, Volume 30, Number 1, pp. 53–67.

form of anonymity in a surveillance-rich high technological environment.

Cover refers to developing a fictional story. Spies use the term in two contexts: 1) *cover for status* becomes your assumed identity. It is your "costume"—say, a repairer, a realtor, job seeker, etc., and 2) *cover for action*, which is the explanation of why you are doing what you are—i.e., sitting in a café checking your e-mails. It should not be confused with the military's use to provide physical protection, as in "provide me with 'covering' fire."

Disguise is an augmentation of physical looks and behaviours to hide a person's true self. Cover and disguise are usually used concurrently to ensure optimal concealment of identity.

Disguise is a useful tradecraft tactic to avoid detection from surveillance by disrupting facial recognition technology. Fundamentally, facial recognition technology relies on three factors: 1) facial detection—identifying the image as a face looking for points such as mouth, eyes, and nose; 2) feature extraction—involving the alignment of the image to normalize the features for better recognition; and, 3) facial recognition—matching the image to a database of images.[50]

Some of the problems associated with successful facial recognition include the inability to identify changes in facial expression to the possible matched image on a

---

50. Shahina Anwarul and Susheela Dahiya, "A Comprehensive Review on Face Recognition Methods and Factors Affecting Facial Recognition Accuracy," in, Pradeep Kumar Singh, Arpan Kumar Kar, Yashwant Singh, Maheshkumar H. Kolekar, Sudeep Tanwar (eds.), *Proceedings of International Conference on Recent Innovations in Computing 2019: Lecture Notes in Electrical Engineering* (Cham: Springer), Volume 597, pp. 495–514.

database and occlusion (the partial covering) of the face.[51] Disguise is a means to exploit those weaknesses in surveillance and facial recognition programs.

There are five elements to developing a disguise. These include blending, changing physical traits, developing different habits and physical movements, carrying props for immediate changes in appearance and "going analog." These five elements have a track record of reducing the likelihood of a person's identity being recognized on surveillance footage and/or through facial recognition technology.[52]

Blending is not standing out in the environment in which you find yourself. It is different from *hiding* as a person is not trying to conceal themselves but rather merge into the "background." Blending in will shield a person's identity by making them unrecognizable and unremarkable. For example, if travelling in the business district of a city, a person would likely start their disguise by wearing business attire or if travelling near a beach dressed in appropriate beachwear. A disguise should not bring attention; instead, it should assist you in mingling.

Changing physical traits is an integral part of disguise. A disguise can make identification more difficult. Changing the color of your hair, hair style, or eye color (through contact lenses) can assist in changing your appearance. But simple changes such as wearing a hat or glasses can alter your physical appearance (but don't wear sunglasses if doing so will make you stand out). Changes in clothing also assist. Carrying a change of clothes

---

51. Loc. cit.

52. Meena N. Harnois, Editor, *Facial Recognition Technology: Best Practices, Future Uses and Privacy Concerns* (Hauppauge, NY: Nova Science Publishers, 2013).

different from the ones you are wearing can also prove effective.

At short notice, hats, sunglasses, umbrellas (and other "props"), and changes of clothes are some of the more immediate measures taken to provide a disguise.[53] They are easily used and then stowed. They are also helpful in confusing facial recognition software programs.

To compliment the change in appearance, adopting different habits and physical movements will provide a further dimension to your disguise. Effective disguises work better when there is a complete change of character to accompany the change of look.

Changing your facial expressions can change your appearance and people's impression of your demeanor. Walking differently, taking longer or shorter strides, or adopting a slower or faster pace will also assist you in developing your disguise.

So too can habits. In disguise, a person may noticeably drink tea instead of coffee; if a person is a non-smoker, they may smoke in places they are likely visible. Essentially, doing visually opposite or unlikely activities for the real identity will help conceal a person's identity.

Disguise must involve changes to your physical appearance and a change in your psychological behavior; that is, developing a new persona that is removed from true yourself but simultaneously able to blend into the place you are in.

---

53.   Raymond P. Siljander, *Fundamentals of Physical Surveillance: A Guide for Uniformed and Plainclothes Personnel* (Springfield, IL: Charles C. Thomas Publisher, 1977).

While not always conscious of it, mobile electronic devices often track our movements and habits.[54] Regardless of the most effective disguise, a personal cell phone will identify you if someone else gains access to that data. While there are some ways to reduce cell phone tracking, say, by a device's privacy settings and adopting encrypted platforms, the most effective countermeasure is to travel *without* a cell phone—"go analog."

The same can be stated for smartwatches or fitness trackers. With these devices, going analog means leaving behind any piece of technology that can be used to track your movements.

> A rule-of-thumb is that if you can track what you're doing yourself, so can someone else. All they need is access to your device's data.
>
> *Troy Whitford, 2022*

Increasingly, national security and law enforcement agencies access the data generated by a person's portable electronic devices. However, to be useful to those agencies, a person must carry one of these devices. But if you do not have such a device, it cannot reveal anything.

Other ways that those who are intent on knowing your movements will use data gathered by you using your credit cards and other cashless transaction cards. Like travelling without a portable electronic device, it is better to pay for items with cash. Without an electronic device and by using cash, you are now "flying under the radar."

---

54. Eric Pait, "Find My Suspect: Tracking People in the Age of Cell Phones," in *Georgetown Law Technology Review*, December 2017, Volume 155.

Criminals and terrorists[55] have long learnt to avoid using electronic devices because they reveal their locations.

Although a disguise will help you move about with less risk of discovery, these props are only secondary to the person's "cover." That is to say, disguise is the outward expression of a cover. Cover is the background, story, or explanation you will use to justify doing what you will be doing or being where you intend to visit. It will protect your true identity.

Developing a cover requires careful consideration and planning. You will need to prepare a story that will *convincingly* explain what you are about to do when you do it. As an example, being dressed in business attire and sitting in the lobby of an office complex would make you blend in. But what if a security officer notices you? What cover story do you give her that will convincingly explain your presence? You might say that you are waiting for a friend to have coffee, a business meeting to begin, or are there to attend a job interview.

All these stories might provide a suitable cover, but greater detail might be needed. The security officer may ask who your friend is, what floor they work on, or which company is conducting the interview or meeting. Subsequently, your cover requires some investigation to gain these simple background facts. In the spy business,

---

55. For instance, see the so-called *Al Qaeda Training Manual*, (the Arabic title being *Declaration of Jihad Against the Country's Tyrants Military Series*) which was discovered in 2005 in the house of Khalid Khaliq, a friend of the 7 July London suicide bomber Mohammad Sidique Khan.

these details are known collectively as *basic intelligence*.[56]

A good cover is a kind that observers will cognitively process themselves. That is to say, anyone noticing you will be led to conclude that you have a plausible reason for being there without questioning. For instance, if a person had you under surveillance in a suburban area, their presence would be noticed by people in the neighborhood. If the surveillant was in a typical sedan, sitting in the driver's seat looking at your house, that person is likely to draw attention. However, with some thought and planning, a cover story and disguise could be developed to allow the surveillant to stay longer than expected.[57] You can reverse these actions to benefit you in your operation.

Driving a pick-up truck dressed in high-visibility work clothes—that of a construction worker, tradesperson, or road maintenance person—provides a cover that allows you to merge with the roadside environment. Carrying a clipboard and appearing to take notes or measurements gives you the appearance that work is being undertaken. If observers can make sense of what you are doing without asking questions, there is less suspicion.

As with the business example, some background preparation is required. In this instance, if asked, you might explain your presence in a different suburb by

---

56. See footnote 23 for a definition of *basic intelligence*. Information might include knowing the businesses that operate at that location and perhaps, some of the names of people that work in the building as well as their roles.

57. Chris Cooper, *Behind the Private Eye: Surveillance Tales and Techniques* (Port Macquarie, NSW: Chris Cooper, 2005).

saying you are assigned to count traffic in the street for potential road improvements.

Creating a cover story requires a few elements. A cover story must be 1) believable and 2) mundane, and 3) does not require any specific expertise or specialized knowledge. Using the previous example of an individual attempting to blend into an office building, you would first have to develop a believable cover story. Yes, a meeting or waiting for someone is a believable explanation, but the meeting or the person who will be met should be unremarkable. The cover story should be of little to no interest to the person inquiring about an individual's presence.[58]

Using a meeting as a cover, the meeting should be simple, like a regular staff meeting or reporting session. The individual would not claim to be a manager or anyone in authority but rather a simple sales representative or administrator. The cover story you create may be detailed, but when you tell it, kept as simple as possible. In this regard, experience shows that less detail is better.

Your cover must not require technical or specific knowledge outside of your area of expertise. For example, you should not claim to be a software engineer if you do not understand the theory and practice of computing programming.

Returning to our example of the "traffic counter," we see that the cover job is mundane, so the surveillant should reflect this perception. The surveillant would also make it clear they are not a civil engineer or city planner but instead a person hired by a temp agency to count traffic.

---

58. Chuck Chambers, *The Private Investigator's Handbook* (New York; Perigee Books, 2005).

The surveillant could then say they don't know what the traffic survey is for because it is a casual job for a day or so.

This type of cover story would be of little interest to the listener, and if the story is delivered well, the surveillant will convey that they have little interest in explaining their role. Further counting traffic takes little to no specialized knowledge, so there will be no expectation that they could provide any expert explanations.

The objective of a disguise and cover is to blend into the surroundings. Remaining unremarkable and unrememberable to others and moving unnoticed is the goal.

Developing a successful disguise and cover requires research and good judgment. As tradecraft, it borrows from the talents of an actor and the creativity of a writer. It begins with developing a keen sense of observation and studying how people move through environments and interact with each other. Once a clear understanding of how people interact and appear in the environment focuses on emulating those people through a similar look and action.

CONCEALMENT AND CAMOUFLAGE

Cover and disguise are used to alter your appearance and provide an explanation about your activities. *Concealment* and *camouflage* differ in that these TTPs allow you or something you to "disappear" rather than "hide in plain sight."[59] Concealment is the use of containers or other apparatuses to avoid detection.

---

59. Van Ritch, *Rural Surveillance: A Cop's Guide to Gathering Evidence in Remote Areas* (Boulder, CO: Paladin Press, 2003).

Figure 4—The color of this lion blends with the environment.

Camouflage uses mimicry to blend into the surrounding environment. This could be using disruptive patterns to blend in with your surroundings or to dress in a way that makes you fit into the environment, as well as to change your outside appearance.

The following illustrates some methods you can use for concealment and camouflage. They also highlight the psychology that concealment and camouflage rely on.

Biologists and zoologists have long studied camouflage in animals. The color of animals blends with their environments—the shape of spots on a leopard or stripes on a tiger are natural camouflage that hides and distorts its view.

The deceptive traits of camouflage used in the animal kingdom rely on the principles of "signal-to-noise ratios," visual perceptions, distortion of surfaces and edges, as well as background complexities.[60]

A signal-to-noise-ratio is where the target of the surveillance is the "signal"—that is, something that can be identified—while the "noise" is anything that interferes with the ability to identify the target. It follows that the greater the noise, the more difficult it is to identify the target.

Visual processing is mentally recognizing what is seen. Effective camouflage interferes with visual processing. The distortion of edges and surfaces also plays a role in visual processing. Surfaces that copy background and edges make it difficult to see where an object begins and ends, making it more challenging to identify.[61]

These animal kingdom camouflage fundamentals are replicated in the military and in nature observers such as ornithologists. By blending colored textures in clothing and equipment with a common background (in the operating environment) or by distorting the object's outline,[62] camouflage has been used by militaries regularly since the nineteenth century and, more recently, by nature viewers.

---

60. Sami Merilaita; Nicholas E. Scott-Samuel; and Innes C. Cuthill, "How Camouflage Works," in *Philosophical Transactions of the Royal Society: Biological Sciences*, 2017, Volume 372, Number 1,724.

61. Ann Elias, *Camouflage Australia: Art, Nature, Science and War* (Sydney: Sydney University Press, 2011).

62. Timothy N. Volonakis; Olivia E. Matthews; Eric Liggins; Roland J. Baddeley; Nicholas E. Scott-Samuel; Innes C. Cuthill, "Camouflage Assessment: Machine and Human," in *Computers in Industry*, 2018, Volume 99, pp 173–182.

Figure 5—British snipers from No. 34 Squadron Royal Air Force Regiment training in ghillie suits in 2015. Photograph by Senior Aircraftman Phil Dye. Courtesy of the Ministry of Defence, United Kingdom.

Military camouflage has evolved from its traditional single khaki color to two, three, or even four colors on a surface. Using a "fragmentation" approach, patterns and shapes were added to uniforms and equipment to further distort edges.[63] The military influence in camouflage on a civilian population is most apparent in naturalists' attire but was even considered a fashion item in the early 2000s.[64]

Aside from nature studies and fashion statements, camouflage applications in the private domains are mostly reserved for private inquiry agents (i.e., private

---

63. Guy Hartcup, *Camouflage: The History of Concealment and Deception in War* (South Yorkshire: Pen & Sword Books, 2008).

64. Eric H. Larson, *Camouflage: Modern International Military Patterns* (South Yorkshire: Pen & Sword Books, 2022).

investigators) conducting surveillance in rural or remote environments. Urban or suburban environments would likely require a disguise or appearance that blends into a dense human-populated landscape.

In a rural or remote environment, there is a lower population density and a person's appearance in such an environment will almost always be noticed or viewed with curiosity by local inhabitants. Subsequently, rural or remote surveillance requires the operative to be invisible rather than blend into a populated environment.

The principles of camouflage apply to conducting surveillance in low-populated areas. The choice of clothing must blend with the given landscape. For example, it would be pointless to dress in green camouflage in a desert-like environment. Using branches, grasses or soil from the landscape provides the most effective camouflage because it is native to the surroundings.[65]

Camouflage netting or ghillie suits[66] are available at most military surplus stores and on-line outdoors retailers. They are often used by bird watchers to cover themselves or their photographic equipment.

The signal-to-noise ratio can be affected by sun reflections from such items as a person's watch or

---

65. Raymond P. Siljander, *Fundamentals of Physical Surveillance: A Guide for Uniformed and Plainclothes Personnel* (Springfield, IL: Charles C. Thomas Publisher, 1977).

66. A ghillie suit refers to a type of camouflage clothing that blends with the wearer's environment, whether it is foliage, snow, or sand. The name is derived from the Scots Gaelic *gille*, which describes a person who works outdoors. David Amerland *The Sniper Mind: Eliminate Fear, Deal with Uncertainty, and Make Better Decisions* (New York: St. Martin's Press, 2017), p. 53.

photographic equipment. To mitigate these issues, a watch can be worn on the inside of the wrist and by using lens caps.

Figure 6—The camera lens can cause reflections that have the potential to be seen many miles away.

Movement often reveals the edges of objects and breaks the continuity between the subject and its background. Sounds in a quiet place also require careful attention and assist in visual processing by identifying a specific location and focus.

Subsequently, effective camouflage will involve a high degree of stillness and silence. The principles of mimicking the environment, merging into the background, and reducing edges are all common approaches. They should be adopted when a citizen who is considering doing something that requires secrecy. Effective camouflage requires a degree of reconnaissance.[67]   It's important to have a familiarity with the terrain.

Figure 7—The late-Don Adams as Maxwell Smart, with famous shoe phone, circa 1968. Courtesy of General Artists Corporation.

Knowing the colors and shades that make up the environment at various times of the day or night makes designing camouflage better informed. Looking for areas with denser backgrounds and practicing camouflage using

67. See, Henry Prunckun, *How to Undertake Surveillance and Reconnaissance: From a Civilian and Military Perspective* (South Yorkshire, UK: Pen & Sword Books, 2015.

different items in areas near (but not at) the location before surveillance allows individuals to test their abilities.[68]

Concealment is the act of hiding an object. There are many reasons why a citizen may want to conceal an object. Often it could be for security purposes or to ensure the object does not bring any attention to itself. Concealment can be undertaken in a few ways. Objects can be hidden in secret compartments or unlikely places such as air vents or electrical outlets.

Objects can be concealed amongst like objects, such as a secret book placed in a library. And objects can be placed within other plain objects like cans of food, or in more covert instances, cameras or even weapons can be concealed on other objects like pens or toys.[69]

The psychology behind concealment is based on how people perceive what is around them and what they expect to see. If they see a can of food or a pen, the mind accepts what is presented, and there is little to no curiosity or thought that it might serve as a shell to conceal another object unless there is a suspicion that they should be looking for something further.

During the Cold War, concealment was regularly used as a spy technique. Some of the more unique concealment devices included shoes with radio transmitters in the heel (*à la* the television show *Get Smart*), lipstick pistols, fake scrotum concealment prosthetics, and dead rats.

---

68. U.S. Marine Corps, *Scouting and Patrolling* (Honolulu, HI: University Press of the Pacific, 2004).

69. M.C. Finn, *The Complete Book of International Smuggling* (Boulder, CO: Paladin Press, 1983).

The Central Intelligence Agency favored using dead rats to conceal notes, microfilm, and money to pass on to agents. Using a dead drop[70] method, a rat's corpse would be gutted and filled with whatever agents wanted to pass on to other agents. The rats were then sprayed with wormwood oil to stop cats from feeding off the carcases.[71] The idea was simple yet effective because no one would give a dead rat in an alley much thought, nor would anyone particularly want to handle it. This feeds onto the earlier discussion of the mind reacting to what it sees. In this instance, a rat carcass is unremarkable but also distasteful and unlikely anyone would interfere with it.

In private investigation and intelligence collection, there have been many developments in how cameras and listening devices are concealed. Cameras have been concealed in wall clocks, pens, neckties, sunglasses, and hats, to name a few. As technology has improved, so has the resolution of the images captured from these devices.

Audio devices that are voice-activated and will even call the surveillant's cell phone have also been developed. Hence, an individual can listen via the telephone to a conversation from anywhere there is a cell phone tower connection.

In law enforcement, there are strict controls on covert photography and the use of listening devices. Any use of these technologies must abide by federal and

---

70. A *dead drop* is a technique that allows information or objects to be passed from one operative to another without making physical contact or being at the dead drop site at the same time. Bob Burton, Top Secret" *A Clandestine Operative's Glossary of Terms* (Boulder, CO: Paladin Press, 1986), p. 37.

71. Declassified information provided by the U.S. Central Intelligence Agency.

state/provincial laws. The principle is that when obtaining information using spycraft devices, the act must not breach a person's expectation of privacy.[72]

By way of example, if a person is sitting in their lounge room with the curtains open or washing their car in their driveway or at a shopping mall, the expectation for privacy is significantly reduced and may likely not break any laws. Concerning listening devices, it's usually unlawful to record a conversation without permission from the subject.

Even law enforcers are bound by laws and need court approval via a judge-ordered warrant to use listening devices. An understanding of the law around covert audio and visual surveillance is essential in your planning process. Legalities aside, concealed cameras are adequate for surveillance activities. Hidden cameras do not bring attention and are discreet.

Concealing documents or other important items requires some imagination. Individuals may construct secret panels or cavities in the walls of their home or hide keys amongst many other no longer used keys (like hiding a book in a shelve of books). Both types of concealment achieve the same outcome but rely on different cognitive perceptions.[73]

---

72. Unless you are party to the conversation, and "the use of the device is reasonably necessary for the protection of the lawful interests of that person." Part 2, Division 1, Section 4(2)(a)(ii) of the South Australia *Surveillance Devices Act, 2016*. Other jurisdictions are likely to have similar legislative provisions. You'll need to get legal advice before using such a device as a citizen.

73. For examples, there are many out-of-print, underground publications that can be found in secondhand books shops and on-

In the case of a wall cavity, the space is hidden and not apparent to the causal viewer. The keys are like hiding an important book in a library with so many that it would be difficult to quickly locate the one the person was looking to find.

Another form of physical concealment is having a decoy. Developing a fake document or item placed where it might be expected to be found is another effective concealment method adopted by those wanting to secure items. For example, a person may stay in a hotel room and want to protect their wallet. It might be possible to place a decoy wallet in the room safe or on the bedside table (where someone might expect to find it) and the actual wallet in another hidden location within the hotel room.

Physical concealment requires a degree of imagination and creativity. Hiding in plain sight relies on using objects and places that are dull or of little interest to others. Hiding objects inside door jams, refrigerators (e.g., dropped into a bottle of orange juice or wrapped in leftovers) and telephones are just some existing items in a hotel room that can be used.

Camouflage and concealment have applications for surveillance operations. The ability to disappear rather than blend has applications in specific settings, while concealing provides the opportunity to collect audio and visual intelligence with little to no suspicion drawn to the citizen operative.

---

line retailers such as, Peter Hjersman, *The Stash Book* (Boulder, CO: Paladin Press, 1978), and Edie the Wire (pseudonym), *How to Bury Your Goods* (Mason, MI: Loompanics Unlimited, 1981).

Figure 8—Spot the hidden USB data drive . . .

It is about eliminating any ambiguities or curiosities people might have and being drawn to further investigation. Camouflage and concealment rely on what psychologists call *motivated perception*, which is the biased, selective ways[74] people view what is around them. Playing to what is expected to be seen is an important part.

---

74. Yuan Chang Leong; Brent L. Hughes; Yiyu Wang; and Jamil Zaki, "Neurocomputational Mechanisms Underlying Motivated Seeing," in *Nature Human Behavior*, 2019, Volume 3, pp. 962–973.

— CHAPTER FIVE —

PREMISE THREE—ANTICIPATING HOSTILE INTENTIONS TO FRUSTRATE THESE SCHEMES

S py novels are awash with examples of undercover agents and their spy handlers meeting to exchange information, money, or instructions. As we know, surveillants are tasked to observe so that they can gather what is termed *actionable intelligence*.[75] The spy methodology common portrayed in espionage stories entails loading or clearing a dead drop or contacting an agent via a brush-past, meeting in a public place, or at a safe house.[76]

To safeguard these exchanges, operatives perform surveillance detection or what is known as *counter-surveillance*. While it is difficult to evade surveillance,[77] counter-surveillance techniques may, at the least, make you aware of surveillance—whether you are being watched—and allow you to alter your plans or seek safety.

---

75. Robert M. Clark, *Intelligence Analysis: A Target-Centric Approach, Second Edition* (Washington, DC: CQ Press, 2007), p. 9.

76. A safe house is a refuge where operatives can meet in secret. Jefferson Mack, *The Safe House: Setting Up and Running Your Own Sanctuary* (Boulder, CO: Paladin Press, 1998).

77. For more detailed information about surveillance methods, see Henry W. Prunckun, *Espionage Black Book Three: Surveillance Explained* (Unley, South Australia: Bibliologica Press, 2021).

COUNTER-SURVEILLANCE

Recently, news stories have described the activities of politically motivated extremists targeting individuals, law enforcement officers, and doctors and patients exercising their civil rights or performing their authorized duties. But anyone can potentially be the target of surveillance—there are plenty of potential adversaries: organized criminal groups, extremists, political activists, malcontents of many descriptions, or government agencies.

As a nation's political landscape changes, what was once legal and permissible may suddenly become illegal and prosecutable. Take, for instance, the Chinese Communist Party's imposed changes on the territory of Hong Kong to abolish democratic representation.[78]

Conversely, those charged with enforcing the law may themselves become the target of extremist surveillance with the intent to intimidate. Following the execution of a court-ordered search warrant on former-president Donald Trump's Palm Beach, Florida residence in relation to his alleged breaches of the *Espionage Act*, "The FBI and DHS have experienced an increase in threats toward federal law enforcement officers and to a lesser extent other law enforcement and government officials…"[79]

In these cases, counter-surveillance TTPs could help reveal surveillance, even if the surveillant(s) cannot

---

78. Clive Hamilton, *Silent Invasion: China's Influence in Australia* (Richmond, Victoria: Hardie Grant Books, 2018).

79. Nicole Sganga, "FBI/DHS Bulletin Warns of 'Increase in Violent Threats Posted on Social Media Against Federal Officials and Facilities'," in *CBS News*. Available at: https://www.cbsnews.com. Accessed September 22, 2022 /miami/news/fbi-dhs-bulletin-mar-a-lago-ohio-gunman-ricky-shiffer-call-to-violence/. Accessed September 19, 2022.

identify themselves. Surveillance detection may involve passive observation or active detection techniques.[80]

Let's first examine passive surveillance. This involves looking for people, vehicles, or devices used in surveillance. Passive surveillance detection requires you to develop a heightened situational awareness and be more mindful of your surroundings and environment. Look for suspicious individuals, strange vehicles parked in your neighbourhood, unusual devices, or occurrences around your home or work environment. Be aware of people seemingly showing undue interest in you and your activities. Also, be mindful of unusual requests to be some new person's "friend" on social media, text messages, and attempts to connect with you through personal or professional sites by people you have never met.

Active surveillance detection uses TTPs to identify surveillance conducted by individuals or a team. For example, you may use a surveillance detection route (SDR) to expose surveillance activities. A surveillance detection route is a planned route through a series of locations designed to elicit a response from surveillance or detect surveillance activities.

An SDR follows a route from one location to another with intermediate stops, timed to establish a set routine. With each turn and stop, the subject looks for vehicles and individuals who may be conducting mobile or static surveillance. Take, for example, a driven route that may include several turns, U-turns, and roundabouts, allowing the driver to observe "tails."[81]   When on foot, the route

---

80. ACM IV Security Services, *Surveillance Countermeasures* (Boulder, CO: Paladin Press, 2005).

81. In the parlance of surveillance, following someone is referred to as "tailing" and the following vehicle is referred to as a tail.

should include natural reverses, such as staircases that weave back and forth, and use reflective surfaces (mirrors, car, and store windows) to detect vehicular and foot surveillance.

Remember that a surveillance detection route is intended only to identify but not evade surveillance. Evasion is rarely successful and attempting to do so increases the perception that the surveillance subject is guilty of some offense. Performing illegal or provocative acts to identify surveillance may get you arrested or cause increased scrutiny—this is not what you want.

Instead, the best tactic is to put the surveillants at ease by keeping routines and engaging in innocuous behaviours. However, once you have identified surveillance, that is the time you need to consider how you avoid exposing sensitive places or activities to protect yourself and people or things you care about.

Detecting digital surveillance requires the use of several software applications as well as a few procedures. These specialized tactics, techniques and procedures will help you identify anomalies in your home network and associated devices by alerting you to surveillance.

Foremost, you need to have a reliable antivirus/anti-malware application installed. There are many commercial programs on the market that are suitable. Using one of these creates a safer, more secure defensive computing environment. These can be combined with the firewall[82] functions built into *Windows* and *Apple* operating systems.

---

82. As in the physical world, a firewall is a security device or software application that monitors the file traffic on a computer network to partition unauthorized traffic. Graeme R. Newman and

Anti-malware applications identify attempts to collect and export your data without your permission. These applications may also identify suspicious processes running in the background. As a rule-of-thumb, you should scan your computer's system periodically to eliminate spyware from your devices.[83]

In addition, consider using a software application such as *CCleaner*,[84] which will remove your browser history, tracking cookies, and identify new software patches and upgrades. This program is also available for cell phones.

The good news in these warnings is that even the best surveillance team makes mistakes. Humans are prone to errors, and their technologies fail, as well as Murphy's Law coming into play at odd times. You should be aware that these mistakes, combined with the additional detection techniques we will describe, force the surveillant(s) to misstep to show their presence.

One indicator that someone is conducting surveillance is called a "demeanor hit." This means that the person conducting surveillance will demonstrate behaviors that are either out of place, out of character, or do not logically fit in with the local population, activities, or environment.[85]

Ronald V. Clark, *Superhighway Robbery: Preventing e-Commerce Crime* (Devon, UK: Willan Publishing, 2003), p. 119.

83. Brian Minick, *Facing Cyber Threats Head On: Protecting Yourself and Your Business* (Lanham, MD: Rowman & Littlefield, 2016).

84. As at this writing, this software application could be downloaded without cost for home users at: https://www.ccleaner.com/ccleaner/download

85. ACM IV Security Services, *Countering Hostile Surveillance* (Boulder, CO: Paladin Press, 2008).

We've all seen executive protection details in news footage where the operatives use concealed radios to communicate with other agents talking into their sleeves or touching their ears. While more advanced devices are less visible and usually voice-activated, observing an individual who appears to be preoccupied with a conversation, seemingly with themselves, and not using a cell phone is a good indicator. Nervousness, inappropriate dress, averting one's gaze, and other behaviours may make you suspect that someone is watching you.

Once, on a trip to Russia, my wife and I were surveilled by a female agent while visiting Novgorod. We know we were being surveilled as we repeatedly saw her in her very conspicuous dress (blue with black polka dots), her repeated presence at various locations throughout the morning. Still, most telling was when she followed my spouse into a restroom, including the stall, excusing herself as she backed out!

Dennis B. Desmond, 2022

For vehicle-based surveillance, be aware of vehicle colors, license plate numbers, driver's appearance, and the number and position of passengers, plus any unique vehicle configurations (antennas, luggage racks, decorations, window tags, etc.). Observing lane positions, changes in speed and direction, and whether vehicles attempt to pass or keep pace with your vehicle may indicate the presence of surveillance.

While behaviours such as repeatedly looking at maps or talking into handheld radios were once good indicators of individuals conducting surveillance, modern technologies such as encrypted cell phones and personal navigation systems make surveillance these behaviours less obvious.

Surveillance indicators also include the characteristics of time, distance, and direction.[86] Time is the frequency and length of time a person or vehicle is detected. Observing the same vehicle nearby during morning commutes may present a false positive as the driver may work nearby. But that vehicle seen during a surveillance detection route or over a day might be an indicator.

> While training in the Washington, DC area, our team ran across another agency's team conducting surveillance. When two surveillance "bubbles" intersect, the result was confusing and disruptive for both teams. The outcome can also be catastrophic for an operational team as it may expose their resources to the target of the surveillance.
>
> *Dennis B. Desmond, 2022*

Distance refers to the length of time and location you observe a vehicle or person in proximity. A vehicle or person that remains in proximity over increasingly longer distances is a possible surveillant. And, lastly, direction. During a normal turn, a change in direction, or a stop into a retail centre, observing a vehicle or person change direction to maintain proximity would be another indicator. Individually, these indicators may be innocuous, but together they may provide proof of surveillance.

It is not just physical surveillance that should be detected; the commercial development of a wide range of spyware and tracking technologies made available to the public also means we may be under threat of digital

---

86. Henry W. Prunckun, *Espionage Black Book Three: Surveillance Explained* (Unley, South Australia: Bibliologica Press, 2021), pp. 24–25.

surveillance. Access to remote access trojans (RATs) and tracking and keylogging software through darknet cryptomarkets[87] demonstrate the ease of access to the public.[88] As a result, digital surveillance has grown more accessible and more pervasive.

Unlike physical surveillance, digital surveillance is much more difficult to detect and often requires a level of technical knowledge that the average person lacks. Whereas physical surveillance relies on the skill, behavior and subtle tradecraft of the individuals conducting the surveillance, the cold, uncaring and constant work of technology is only belied by its failure or foreknowledge.

You should be aware of any changes to your system's performance. Is your computer running more slowly than usual? Are there programs or windows that randomly start or stop functioning? Are there any unexplained outgoing e-mails or transfers of data from their device? You can check by examining daily bandwidth usage through your ISP (Internet service provider).

Are there any unrecognized devices connected to your home network? You can check by examining your network's administrative dashboard on-line and your computer's Wi-Fi settings.[89] And have you noticed any technicians working around your property, or have you

---

87. Kaleigh .E. Aucoin, "The Spider's Parlor: Government Malware on the Dark Web," in *Hastings Law Journal*, 2018, Volume 69, p. 1,433.

88. Ron Deibert, "Protecting Society from Surveillance Spyware," in *Issues in Science and Technology*, 2022, pp. 15–17.

89. Mary Pat McCarthy and Stuart Campbell with Rob Brownstein, *Security Transformation: Digital Defense Strategies to Protect Your Company's Reputation and Market Share* (New York: McGraw-Hill, 2001).

left your devices unattended during a service technician visit?

While not always nefarious, it is always good to know what is happening around and, in your home, and never leave your electronic communications devices unsecured.

You may also consider taking preventative measures designed to *disrupt*. However, there are some other interrupting measures that an individual may take to disrupt surveillance effectiveness further.

In the case of a government surveillance team, they will gather as much information as possible to become intimately familiar with you—their target. Their intelligence analysts tasked to support the surveillants will access the Internet, and more specifically, your social media, to develop a pattern of life through which the surveillance team may create their plan.

Surveillance planners can estimate the size of the team needed, the type of surveillance (foot, vehicular, or air), and the locations required to be covered based on your activities, hobbies, social contacts, and locations such as your home, work, family, and recreation. Photographs of you and your family, favorite restaurants, and home or work location also save the planning team a lot of effort.

So, the first step is to "scrub" your social media and/or create alternative accounts under aliases that are not tied to your actual identity.

It is safe to assume that most apps on your smart phone collect information such as geolocation data and personal activity data.[90] These data can reveal a lot about you; your

---

90. Applications on your mobile devices can collect and report your personal data, while ad servers will be able to track your

movements and intentions can be deduced if the device is seized or subpoenaed. The best way to avoid giving away where you have been and for how long is to leave personal electronic devices at home.

But this may not always be possible, so consider privacy precautions. While turning off location services and disabling Wi-Fi and *Bluetooth* are helpful, your phone or tablet will continue communicating with nearby towers for billing and data management purposes. If you are going to a location that you do not what others to know about, consider purchasing a temporary (i.e., "burner") phone and SIM card with *cash* or a stored value card (aka "cash card" or a "gift card").

To avoid tracking devices, keep your vehicle secured, preferably in a locked garage. Where that is not possible, the placement of cameras to monitor your property and access to your vehicle will help. Before using your vehicle, perform a visual inspection or walk around to look for broken taillights, tracking devices, painted marks or antennas. These tactics are used to help surveillance identify your car in traffic. Above all, DO NOT use an employer-provided vehicle for obvious reasons.

When you park in a parking lot, park the vehicle in the open, away from other vehicles, where you can see if someone attempts to place a tracking device or mark the vehicle. Phones can often be used to identify attached *Apple* air tags or *Bluetooth*, or Wi-Fi-connected devices. Use these services if you have them.

If renting a vehicle, which is preferable if you are driving to or near a sensitive location, never sync your

---

location. Kevin D. Murray, *Is My Cell Phone Bugged?* (Austin, TX: Emerald Book Company, 2011).

phone to the car's audio system because this will leave a record of you being present and could also expose your contacts. Never use a rental car's navigation system; instead, rely on a cell phone with a navigation system or a physical map. Ask the rental car company if they use tracking devices that monitor location, speed, and fuel fill-ups. Also, be aware of *On Star*[91] or similar services with their built-in monitoring system. If any of these are used, find another car rental company that does not.

Stored value cards, gift cards or cash cards purchased with *cash* instead of credit or debit cards should be used when making purchases related to the activities you want to be kept private. This tactic applies to purchases at the so-called brick-and-mortar retailers and on-line sellers.

To access on-line retail sites securely, consider installing a virtual private network (VPN) application because you will no doubt be using the World-Wide Web to research activities—like car rental firms. A VPN does not show your true location—it might show that you are in Costa Rica, or Iceland, or other places around the world. Thus, if a local law enforcement agency obtained a search warrant or issued you a subpoena to your ISP, they would find nothing because the service you used to access the World Wide Web was in another country.[92]

---

91. *On Star* , which is marketed by other names around the world, is a subscription-based proprietary in-vehicle communication, security, emergency, and navigation system. Given this description, it is obvious the amount of information an *On Star* system can divulge.

92 Fergus O'Sullivan, "What Do VPNs Share with Law Enforcement?" Available at: *How to Geek*: https://www.howtogeek.com/787544/what-do-vpns-share-with-law-enforcement/. Accessed date September 19, 2022.

Alternatively, using *The Onion Router* (Tor) as your browser will add a layer of security by routing your searches and browser readings through other countries the way a VPN does. This will conceal both the source and the destination Internet Protocol (IP) address, which law enforcers would use to prosecute you.[93]

Your browser history and activities will be concealed using the product *CCleaner*, especially if you combine this free application with the private browsing mode built into Mozilla's *Firefox* browser. There are also various browser extensions or plugins you may use to improve your on-line privacy.

When combined with these surveillance detection measures, surveillance countermeasures will improve your safety and security and provide peace of mind. These techniques reduce your likelihood of criminally motivated identity theft, fraud, and financial loss.

---

93. Besides law enforcers, business insiders who have access to your account information, and activities may provide these data to criminals or extremist activists.

— CHAPTER SIX —

PREMISE FOUR—USING THE SIMPLEST AND LEAST INTRUSIVE MEANS POSSIBLE

T he rise of extremist ideologies has resulted in increased threats of violence to law enforcement, government workers, and citizens with commonsense viewpoints. As such, the ease with which your personal information can be collected, both physically and digitally, requires us to be more active in managing our data. Spy tradecraft is an effective way to protect our privacy, improve our safety and security, and mitigate the harm posed by others to those for whom we care.

While most of us would like to think that we will never have to worry about coming under surveillance by law enforcers, there may come a day when we do. For example, the Arizona law enacted in September 2022 making it illegal to photograph police officers from closer than eight feet. This law also criminalized the filming of police on private property even if the property owner gave their permission.

The law was challenged in the U.S. Federal District Court and was struck down by the judge as a violation of a citizen's First Amendment right.[94] In this context,

---

94. The Associated Press, "Federal Judge Blocks Arizona Law Limiting Filming of Police," *NBC News*. Available at: https://www.nbcnews.com/news/us-news/federal-judge-blocks-arizona-

reflect on the case of Mr George Floyd, where bystanders' cellphone videos were credited mainly with showing the misconduct of a few Minneapolis police officers in the death of Mr Floyd. Suppose the Federal District Court judge had not pointed out the unconstitutional nature of the Arizona law. In that case, any of us could have been arrested for doing what we'd argued to be our civic and moral duty—filming an abuse of state power.

These are not, by no means, the only situations where a law-abiding citizen may be driven to ignore an illegal or immoral law. Take the case of *Roe v Wade*, which stood from 1972 to 2022 as the Constitutional basis for women to decide their health needs when it came to ending a pregnancy. The ruling was overturned in what has been argued by legal scholars to be a *political* decision rather than a *legal* one,[95] resulting in some American states outlawing the medical procedure.

To underscore the political nature of this questionable court decision, President Joe Biden signed an Executive Order to prevent federal agencies and law enforcers from cooperating with state authorities that sought to prosecute

---

law-limiting-filming-police-rcna47148. Accessed September 11, 2022.

95. U.S. Supreme Court Justice Elena Kagan stated, "When courts become extensions of the political process, when people see them as extensions of the political process, when people see them as just trying to impose personal preferences on society, irrespective of the law, that's when there's a problem." Cited in Lawrence Hurley, "Justices Join Debate on Supreme Court's Legitimacy After Abortion Ruling," in *NBC News*. Available at: https://www.nbcnews.com/politics/supreme-court/justices-join-debate-supreme-courts-legitimacy-abortion-ruling-rcna47795. Accessed: September 19, 2022.

women for seeking terminations in neighboring states.[96] In part, he said, "Fundamental rights—to privacy, autonomy, freedom, and equality—have been denied to millions of women across the country, with grave implications for their health, lives, and wellbeing. This ruling will disproportionately affect women of color, low-income women, and rural women."[97]

PREMISE ONE—SECRECY IN PLANNING

Let's suppose that there is a woman living in a state or province where medicinal cannabis oil is illegal, but she needs it to treat her teenager who has severe epilepsy or a parent with Alzheimer's.

What security issues would she need to consider if she wanted to visit a nearby state to obtain this therapeutic treatment? You can substitute your own situation with this one, but any example will demonstrate the principles that apply to other circumstances where injustices exist, yet citizens need to act.[98]

---

96. Known as, *Executive Order 14076* (Protecting Access to Reproductive Health Care Services), signed July 8, 2022.

97. The White House, *Fact Sheet: President Biden to Sign Executive Order Protecting Access to Reproductive Health Care Services* (Washington, DC: White House Briefing Room, July 8, 2022).

98. If you are considering "whistle blowing," do not do what Edward Snowden did; you need to follow the legislative guidelines for bringing illegal maters to the attention of authorities. So-called whistle-blower laws provide protection for people who expose such matters. What Snowden did has been argued to be treasonous. See, Ed Morrissey, "Edward Snowden Broke the Law and Should be Prosecuted," in *The New York Times*, December 18, 2013. Available at: https://www.nytimes.com/roomfordebate/2013/06/11/in-nsa-leak-case-a-whistle-blower-or-a-criminal/

The first step is to revisit Premise One—secrecy in planning. Recall from our discussion in Chapter Three that several issues must be considered.

The first is that even private communications are not private. There is always the likelihood that when a message is transmitted from the sender to the receiver—whether it is oral, physical, or electronic—it can be overheard or read. The chance that someone would be interested in knowing your business is a product of *likelihood* and *consequence*. This ten needs to be viewed in the context of *threat*—who might be interested in what you are doing (revisit our discussion on threat levels in Chapter Three).

If you are planning an outing with a friend to the zoo, no one is likely to be interested. But, if you were, say, a Hong Kong citizen planning to attend a pro-democracy rally, you can be certain that the Communist Chinese Party's intelligence service will be interested.

This brings us to the issue of pre-, peri- and post-event communication. If you take little or no security measures in the before and during event stages but apply security TTPs after the event, you may think you are safe having attended the pro-democratic rally. But what you have done is left a trail of communications that investigators can retrieve and use to prosecute you.

The lesson is that if you are going to challenge an illegal or unconstitutional law, you must implement communications security at the point where you consider doing it.

---

edward-snowden-broke-the-law-and-should-be-prosecuted. Accessed September 27, 2022.

Stop and Think

Here are several "seed" ideas for you to consider. The aim is to help you broaden your thinking to include other areas of communication that you use so you can lay out a more comprehensive security plan. Remember, if you plan for the worst situation—i.e., surveillance by a state intelligence agency—then you will be protected against lesser "threat agents"[99]—con-artists, petty criminals, and society's other miscreants.

Person-to-Person: Make a list of how you will ensure that others will not overhear you discussing your planning. Consider events such as speaking on the phone or talking to a friend, colleague, or family member in person.

_____

_____

_____

_____

Correspondence: By correspondence, we are referring to any written communications. Letters mailed through the postal service, notes delivered by hand, or e-mails sent via the Internet. Can you think of other forms of correspondence you'd use when making your plans? Social media comes immediately to mind, as do text messages. Against each method, consider how you would

---

99. A "threat agent" is a person(s) or organization that has *intent* and *capability* to cause harm. Intent comprises *desire* and *expectation*, while capability comprises *knowledge* and *resources*. Hank Prunckun, *Methods of Inquiry for Intelligence Analysis, Third Edition* (Lanham, MD: Rowman & Littlefield, 2019), pp. 181–182.

reduce the chance of that message being read before, during, or after your planned event.

_____

_____

_____

_____

Digital Voice and Data: This is another broad category of communication means. As such, you need to think through all the devices you use to search for, read, store, create and amend information using digital communication devices.

The obvious device is the smartphone It allows for voice communications and the sending of photographs and other files, and of course, will leave a record of the date, time, and the size of the data file sent or received as well as how long a voice communication lasted and from whom it was exchanged and the location you were.

As handy as smartphones are, we can see that if you use one, those people interested in what you are or have done will delight in what these data tell them.

But what about your desktop computer, laptop, tablet, or other electronic gizmos? Have you considered them? In particular, your work computer? Because work devices (including smartphones, tablets, e-calendars, etc.) are not yours to "scrub," the rule-of-thumb is DO NOT use a work device. As convenient as it is to use one of these for your personal needs, DO NOT use a work computer for ANY type for private Web searches, on-line purchases, or sending or receiving e-mails, etc. DON'T! DON'T! DON'T![100]

Cataloge your digital devices, then consider how you will safeguard each from others accessing the logs and other digital trails created while you search the World Wide Web or send and receive e-messages.

_____

_____

_____

_____

_____

PREMISE TWO—MOVING SURREPTITIOUSLY

Once planned, your scheme needs to be implemented. This means moving from where you are to where you want to be.

Chapter Four discussed cover, disguise, concealment, and camouflage. Recall that cover was the fictitious reason for being somewhere, and disguise was what was used to alter your appearance so people could not recognize you. Concealment is hiding or the art of making someone disappear. Camouflage is the technique that allows you to blend in with your surroundings.

Moving to an event, being at the event, and returning to your home or another place of retreat can be secured by employing these four techniques. Looking at the three phases of an event—before, during, and after—will help

---

100. Employer provided computers are also known to have what is called *bossware* installed. These are software applications that enable managers to keep tabs on what workers do.

you think through the different TTPs you can employ at each stage.

Stop and Think

Pre-Event: What does your plan say about getting to the event? Is it by public transport (e.g., bus, train, or subway), or private vehicle (your car or transported by a friend)? If you think ahead, will there be any traffic video cameras that will record your vehicle's movement? Train and subway platforms also have cameras. On board buses, trains, and subway carriages, you may also encounter fixed video cameras.

You'll need to consider how you dress to blend in with those around you and/or disguise yourself so no after-the-fact inquiry can determine it is you. Brainstorm several disguises you could use. Consider using one disguise to the event and another on your return travel. Also, consider how you can hide from video cameras, such as standing behind other people or around the corner of the camera or using a COVID-19 protective mask.

_____

_____

_____

_____

Peri-Event: Without a specific event example, using cover, disguise, concealment, and camouflage are difficult because each circumstance is different. Nevertheless, when you consider your event, think in terms of these techniques—what cover story will you use? What disguise will you don (remembering it can be different to the one you wore to the event)? Can you conceal yourself

67

while attending? And is there a way of blending into the surroundings?

_____

_____

_____

_____

Post-Event: This is the "getaway." Once your mission has been accomplished, what is your method of return travel? As you did in the before stage, the same issues need to be considered. If you use the same type of transport, the same issues will no doubt arise, so as long as you consider these, you should be fine. But if you use a different mode or a different route, then be sure you think about all the possibilities inquiry agents will use to detect what you have done.

_____

_____

_____

_____

PREMISE THREE—ANTICIPATING HOSTILE
INTENTIONS TO FRUSTRATE THESE SCHEMES

This is the phase where you need to keep your wits about you. This is because this phase may—depending on what you are planning, may place you under direct surveillance with those out to thwart you. We discussed counter-surveillance in Chapter Five.

In the real spy-world, counter-surveillance TTPs are used to identify, disrupt, degrade, or, in some cases, help avoid hostile surveillance. Counter-surveillance teams use these tactics, techniques, and procedures to selectively limit the information the target exposes about themselves in their daily comings-and-goings. To citizens, planning an operation, counter-surveillance TTPs can help improve your security and afford some control over your information exposure. When you know you are being surveilled, you are better prepared to alter your behavior.

Stop and Think

The subject literature on spy tradecraft makes it clear that surveillants use remote access applications, location tracking software, and open-source information from social media platforms. Nonetheless, you can limit the information your challengers collect.

Consequently, it is important to consider how you will "wipe" your trail of search terms, the Web pages you visited, phone numbers you called, and e-messages you sent and received, and to ensure the information in the public domain is harmless. On this point, some jurisdictions have "data retention" laws. These laws require telephone and Internet providers to retain users' metadata for many months. The rationale is for investigators to access these data in the case of a terrorist event. But misused, these data can also expose your operation.

Covering Your Tracks: Research the most reputable disk and memory cleaners. Start by referring to the programs mentioned in Chapter Five. 1) Download what you consider the best fit for the device(s) and run the application to ensure that you can do this task when you embark on your mission. 2) Research *The Onion Router* (Tor) to understand how it works and why it will help you circumvent creating e-trails. Then, download the program and run it. Use it for all matters directly or indirectly associated with your potential mission. If you prefer, research the use of a VPN.

_____

_____

_____

_____

Laying a False Trail: Social media is fun. Sharing information about your life and posting photographs can be rewarding until you realize that the information you have placed on these websites can be used against you.

Prosecutors and develop timelines of where you were, what you did, and what you thought by the information you voluntarily put in the public domain. Oh, you say that your "privacy settings" are set to "high." That is good, but it is no match to a court order or a law enforcer's warrant. Remember, some jurisdictions have data retention laws.

Search the Web for your name and any variants of it— i.e., if you have a shortened name, a nickname, you use your middle name, or your last name is often misspelled, etc. Why do this? Because your adversary will do the same.

What have you found? Anything—we repeat— *anything* that comes close to showing or expressing an interest in doing what you plan to do or a dislike for what you are about to demonstrate against (metaphorically or otherwise), remove it. If you don't, "Anything you say may be used against you in a court of law."

_____

_____

_____

Situational Awareness: This is a term that means you can perceive the things and events in the environment in which you find yourself. Moreover, you can understand the meaning of these things and events and understand what they might mean in future.[101] What does this psycho-social theory have to do with frustrating hostile intentions? Quite a lot.

If an adversary wants to know what you are doing—because you have "scrubbed" all information about your thinking from the Web—they will have to revert to following you. Physical surveillance! If you have developed situational awareness, you are more likely to take note of individuals or vehicles travelling nearby, travelling in the same direction, or showing undue interest in your activities.

Recall from our discussion about how cover, concealment, camouflage, and disguise (Chapter Four—Premise Two, Moving Surreptitiously) spy tactics, techniques, and procedures can help you to become "invisible," well, that is what your adversary will be doing also. If you have cast your mind across how you would move through the environment undetected, reverse engineer your illusionary thought processes to list a few surveillance indicators.

_____

_____

_____

---

101. Micra R. Endsley, "Toward a Theory of Situation Awareness in Dynamic Systems," in *Human Factors: The Journal of Human Factors and Ergonomics Society*, 1995, Volume 37, Number 1, pp. 32–64.

A FINAL THOUGHT

Using our academic training and subject knowledge, we have separated fact from fiction to produce this technical monograph. We have also developed a theory comprising of four premises that place this real-world information in a practical context: 1) secrecy in planning; 2) being able to move surreptitiously; 3) anticipating hostile intentions to be able to frustrate these schemes; and 4) accomplishing your plans by the "simplest and least intrusive means possible."

Therefore, whether you plan to travel overseas, live in troubled lands, avoid local sociopaths or global criminals, or simply want to live a quiet life, we hope our sensible approach to providing moral, honest citizens with an understanding of spy tradecraft assists.

- o O o -

ABOUT THE AUTHORS

Dr Dennis B. Desmond, BA, MA, PhD, is a Lecturer in cyber intelligence with the School of Science, Technology, and Engineering, University of the Sunshine Coast, Queensland. He is a former U.S. Army Special Agent, an FBI Special Agent, and a Special Agent with the U.S. Defense Intelligence Agency.

Dr Troy Whitford, BA, MA, PhD, is a Senior Lecturer in intelligence and security studies with the Australian Graduate School of Policing and Security, Charles Sturt University, Canberra. He is a former private intelligence operative who conducted intelligence-led investigations in the corporate, political, and legal sectors.

Dr Henry (Hank) Prunckun, BSc, MSocSc, MPhil, PhD, is an Adjunct Associate Research Professor in intelligence methodologies with the Australian Graduate School of Policing and Security, Charles Sturt University, Sydney. He is a former Australian government intelligence analyst. He spent much of his twenty-eight-year operational career in tactical intelligence and strategic research but also served operationally in security, investigation, and counterterrorism.

INDEX